Eagles

by Adele D. Richardson

Consultant:
Tanya Dewey, PhD
University of Michigan Museum of Zoology
Ann Arbor, Michigan

CAPSTONE PRESS
a capstone imprint

First Facts is published by Capstone Press,
1710 Roe Crest Drive, North Mankato, Minnesota 56003.
www.capstonepub.com

Library of Congress Cataloging-in-Publication Data
Richardson, Adele, 1966–
  Eagles / by Adele D. Richardson
    p. cm.—(First facts. Birds.)
  Includes bibliographical references and index.
  Summary: "Discusses eagles, including their physical features, habitat, range,
and life cycle"—Provided by publisher.
  ISBN 978-1-4296-8683-9 (library binding)
  ISBN 978-1-62065-250-3 (ebook PDF)
  I. Title.
  QL696.F32R5353 2013
  598.9'42—dc23
                              2012007511

Editorial Credits:
Lori Shores, editor; Juliette Peters, designer; Kathy McColley, production specialist

Photo Credits:
Alamy: All Canada Photos/Grambo Photography, 17, James Campbell Wildlife, 19, Paul
Harris, 13, SuperStock, 15, Visual&Written SL, 20; Corbis: Daniel J. Cox, 18, Kevin Schafer,
11; Getty Images: Stockbyte, 5; Shutterstock: FloridaStock, 1, Iv Nikolny, 7, Jeffrey Ong
Guo Xiong, 6, Lori Martin, 12, Mike Truchon, 21, viseralimage, 8, worldswildlifewonders,
cover

Artistic Effects
Shutterstock: ethylalkohol, pinare

Essential content terms are **bold** and are defined at the bottom of the page where they
first appear.

Printed in the United States of America in North Mankato, Minnesota.

042012   006682CGF12

Table of Contents

Birds of Prey

Eagles seem like daredevils as they dive from the sky. But these large birds are masters at hunting. Eagles have good eyesight for spotting animals from far away. They use their sharp **talons** to grab and carry **prey**.

Eagle Fact!

An eagle's eyesight is four times better than a human's. Eagles can see as far away as 1 mile (1.6 kilometers).

talon—a large, sharp claw

prey—an animal hunted by another animal for food

wings

eyes

beak

bald eagle

talons

5

Big Birds

Eagles are large birds. They have a **wingspan** of up to 8 feet (2.4 meters). Most eagles weigh between 7 and 12 pounds (3.2 and 5.4 kilograms). The largest eagles weigh as much as 20 pounds (9 kg).

white-bellied eagle

Eagle Fact!
White-bellied sea eagles eat fish and sea snakes.

wingspan—the distance between the outer tips of a bird's wings

African fish eagle

Most eagles have black or dark brown feathers. Some eagles have white or golden feathers on their heads and shoulders.

bald eagle

Hungry Eagles

Eagles eat small animals such as rabbits, squirrels, fish, and birds. They also eat large animals such as monkeys and small sheep. Bumps on their feet help them hold slippery fish. They tear the meat with their hooked beaks.

Eagle Fact!

Bald eagles are not bald. They have white feathers on their heads. These large eagles have nearly 7,000 feathers on their bodies.

Eagles at Home

About 60 kinds of eagles live around the world. The only place eagles don't live is in Antarctica. Golden eagles and bald eagles live in the United States.

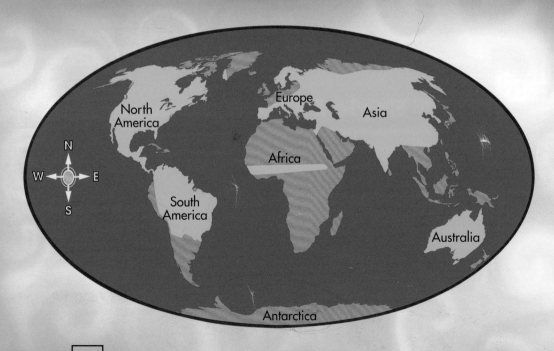

North America

Europe

Asia

Africa

South America

Australia

Antarctica

N
W • E
S

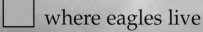 where eagles live

harpy eagle

Eagles live in many **habitats**. Harpy eagles live in **rain forests**. Bald eagles build their nests in trees near water. Other eagles live in jungles, forests, or deserts.

habitat—the natural place and conditions in which a plant or animal lives

rain forest—a thick forest where a great deal of rain falls

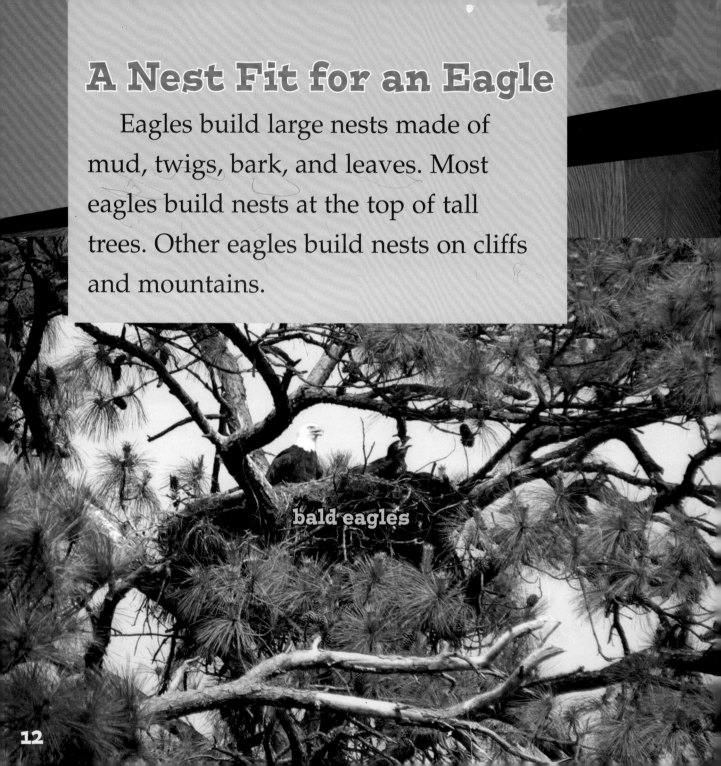

A Nest Fit for an Eagle

Eagles build large nests made of mud, twigs, bark, and leaves. Most eagles build nests at the top of tall trees. Other eagles build nests on cliffs and mountains.

bald eagles

Eagle Fact!

Large eagle nests weigh up to 6,000 pounds (2,722 kg).

short-toed eagle

Eagles use the same nest every year. They add new materials each year. A new nest may be only 1.5 feet (0.5 m) deep. Old nests are huge at 6 to 20 feet (1.8 to 6 m) deep. They can be 10 feet (3 m) across.

Eagle Families

Most eagles **mate** in early spring. The female lays one to three eggs. She stays in the nest to protect the eggs and keep them warm. The male eagle brings food to the nest for her.

Eagle Fact!

Many eagles, including bald eagles, mate for life. They will stay together until one dies.

mate—to join together to produce young

bald eagle with eggs

Eaglets

Newly **hatched** eagles are called eaglets. They are never left alone for long. Parent eagles tear up small pieces of food for the eaglets.

Light gray **down** covers an eaglet's body. Regular feathers grow after two to three weeks. It takes 10 to 13 weeks for all the adult feathers to grow. Then the young eagles learn to fly and hunt.

hatch—to break out from an egg

down—soft, fluffy feathers of a bird

Life Cycle of an Eagle

Newborn: Eaglets cannot tear up their own food until they are 5 to 7 weeks old.

eaglets

Young: Young eagles stay near the nest until they can hunt and feed themselves.

Adult: Eagles in the wild live 10 to 30 years. Some eagles live up to 50 years in zoos.

Danger!

Eagles have very few **predators**. Raccoons, snakes, and great horned owls sometimes eat eagle eggs and eaglets. Parent eagles fiercely guard their young.

golden eagles

tawny eagle

Eagle Fact!
Eagles are careful to avoid danger. They stay away from people. Eagles will attack animals and people to defend their young.

People sometimes harm eagles. At one time, people hunted and killed some kinds of eagles until they almost became **extinct.**

predator—an animal that hunts other animals for food

extinct—no longer living anywhere in the world

People Helping Eagles

People work to help eagles survive. Some people treat sick eagles and return them to the wild. People have passed laws against killing eagles. Laws also protect eagle habitats.

golden eagle

Amazing but True!

The bald eagle is a strong flier and powerful diver. Bald eagles can fly at speeds of 30 to 35 miles (48 to 56 km) per hour. When diving for prey, bald eagles can reach speeds of 75 to 99 miles (121 to 159 km) per hour. The strike from an eagle's talon at that speed is stronger than a rifle bullet.

Glossary

down (DOUN)—soft, fluffy feathers of a bird

extinct (ik-STINGKT)—no longer living; an extinct animal is one that has died out, with no more of its kind

habitat (HAB-uh-tat)—the natural place and conditions in which a plant or animal lives

hatch (HACH)—to break out of an egg

mate (MATE)—to join together to produce young

predator (PRED-uh-tur)—an animal that hunts other animals for food

prey (PRAY)—an animal hunted by another animal for food

rain forest (RAYN FOR-ist)—a thick forest where a great deal of rain falls

talon (TAL-uhn)—a large, sharp claw

wingspan (WING-span)—the distance between the outer tips of a bird's wings

Read More

Dolbear, Emily J. *Bald Eagles.* Nature's Children. New York: Children's Press, 2012.

Lundgren, Julie K. *Eagles.* Raptors. Vero Beach, Fla.: Rourke, 2010.

Miller, Sara Swan. *Eagles.* Paws and Claws. New York: PowerKids Press, 2008.

Internet Sites

FactHound offers a safe, fun way to find Internet sites related to this book. All of the sites on FactHound have been researched by our staff.

Here's all you do:

Visit *www.facthound.com*

Type in this code: 9781429686839

Super-cool stuff!

Check out projects, games and lots more at
www.capstonekids.com

Index